Breeding Rhode Island Reds For Type and Egg Production

by H.A. Bittenbender

with an introduction by Jackson Chambers

This work contains material that was originally published in 1922.

This publication is within the Public Domain.

This edition is reprinted for educational purposes
and in accordance with all applicable Federal Laws.

Introduction Copyright 2017 by Jackson Chambers

Self Reliance Books

Get more historic titles on animal and stock breeding, gardening and old fashioned skills by visiting us at:

http://selfreliancebooks.blogspot.com/

Introduction

I am pleased to present yet another title on Poultry.

The work is in the Public Domain and is re-printed here in accordance with Federal Laws.

As with all reprinted books of this age that are intended to perfectly reproduce the original edition, considerable pains and effort had to be undertaken to correct fading and sometimes outright damage to existing proofs of this title. At times, this task is quite monumental, requiring an almost total "rebuilding" of some pages from digital proofs of multiple copies. Despite this, imperfections still sometimes exist in the final proof and may detract from the visual appearance of the text.

I hope you enjoy reading this book as much as I enjoyed making it available to readers again.

Jackson Chambers

Fig. 1. Hen No. 346, a low producer. First year, 54 eggs; second, 76; first half third, 22.

Fig. 2. Hen No. 345, shallow breasted, lacking in depth, medium in length of back, but with more width thru back, body and breast than fig. 1.

Fig. 3. Head of cockerel, showing breeding strength and masculinity.

Fig. 4. Head of male lacking in masculinity and breeding strength.

BREEDING RHODE ISLAND REDS FOR STANDARD TYPE AND EGG PRODUCTION

BY H. A. BITTENBENDER

In the past ten years the Rhode Island Red has become a very popular farm fowl thruout the cornbelt states, and particularly in Iowa. Evidently it seems to meet farm conditions, but at the same time criticisms have been made concerning the maintenance of type and egg production which led the Poultry Section of the Iowa Agricultural Experiment Station to make a study of breeding methods with Rhode Island Reds. The results show how this popular fowl may be kept up very satisfactorily in type and production by due care in selection and breeding.

The main criticisms of the Rhode Island Red that have come to our attention are these:

First: That it is particularly difficult to maintain a flock of Rhode Island Reds uniform and true to standard type.

Second: That it is extremely difficult to breed and keep the dark, even red color and that the hens lose their red color and become lighter when they molt.

Third: That Rhode Island Reds are lazy, have a tendency to become over-fat and are poor producers.

Fourth: That Rhode Island Reds are more inclined to be broody than other varieties.

To discover breeding methods that would be helpful in maintaining the characteristics of the Rhode Island Reds, we used the experimental flock of Rhode Island Reds and secured additional birds from leading breeders thruout the country. A large number of matings were made. The birds were all trap-nested and the chicks pedigreed. Trap nesting with the experimental flock of Rhode Island Reds began in 1908 and breeding records have been kept since that time. No particular attention was paid to selection for standard disqualifications and breeding defects until 1917, but from 1917 on, particular attention has been paid to Rhode Island Red breed type and color as well as egg production.

STANDARD TYPE

The breed type of the Rhode Island Red, as established by the American Standard of Perfection, has been studied very closely. The ideal type is of long and rectangular build. In the experimental flock several distinct rectangular types were found. The rectangular type cannot be judged entirely from a side view. A bird may show the long, flat back, the square breast and body and the square under-line, and thus give the rectangular type from a side view, but it might not constitute a good type. It is equally essential that the bird must have width thru the chest and body with the rectangular shape.

Fig. 1 shows a bird quite closely approaching the standard type, particularly in the back, in that it is long and flat. This hen also has a long body, but is lacking in breast and depth of body. Her worst defect is that she is narrow thruout. At the first glance she seems to be a good type, but because of the lack of width she does not have the constitution, vigor, vitality, or the capacity for good egg production. Her record is very low.

Fig. 5. Hen No. 9631. First year's egg production, 188; second, 154.

The hens in figs. 1 and 2 (Hen No. 346 and Hen No. 345) were hatched in the same incubator, brooded in the same brooder, reared and handled under exactly the same conditions. Compare the type of Hens No. 345 and No. 346 with fig. 5 and note the differences in essential character. Fig. 5 shows a hen of the true rectangular type, long back, full breast, and deep body. Not only is the hen rectangular from a side view, but she has good width which carries uniformly from the wings to the tail, and she is a good producer.

In selecting Rhode Island Reds it is absolutely essential that the rectangular type be kept in mind, but in selecting for the long body be sure that the individual has a full breast and deep body and uniform width.

CONSTITUTIONAL VIGOR

Both the male and the female should be selected for constitutional vigor and vitality. Vitality, type and egg production are very closely correlated.

In selecting males for vigor, particular attention should be paid to masculinity and maturity. Fig. 3 shows the head of a cockerel with excellent vigor, vitality and masculinity, as compared with fig. 4, a male lacking in both vitality and masculinity. The particular difference between these two birds can be seen in the prominence of the eye, the width, depth and breadth of head, shortness and stoutness of the beak of the bird shown in fig. 3 as compared with fig. 4, a cockerel with a longer, flatter head and sunken eye. The only difference in these two birds is parentage, as they were reared under identical conditions and had the same opportunity for development.

Physical Strength	*Physical Weaknesses*
HEAD — Short, broad and deep.	HEAD — Long, narrow and lacking depth from top to base of beak; nostrils small and elongated.
BEAK — Short, stout, broad and well curved.	BEAK — Long, straight and pointed.
EYES — Bright, alert and prominent.	EYES — Dull and sunken.
COMB — Red in color, well developed in size.	COMB — Undeveloped and often pale.
BODY — Broad, uniform in width, especially across the back.	BODY — Narrow, especially thru back; lacking in depth.
BREAST — Full and well developed.	BREAST — Undeveloped and sharp.
LEGS — Stout, placed directly beneath the bird, knee or hock joints wide apart.	LEGS — Long and stilt-like or bending at hocks, giving the bird a squatting appearance.
TOES — Straight and toe nails well worn.	TOES — Long, toe nails sharp.

The breast in the body conformation of the two birds shown in fig. 6 and fig. 7 is very square. The slower maturing, longer geared males rarely make as good breeders as those males that mature earlier, providing they reach standard weight.

In a general way, the characteristics which indicate breeding strength in males also apply to hens. Fig. 8 shows a pullet of excellent Rhode Island Red type, combined with constitutional vigor and vitality, while fig. 9 shows a pullet of the same age, reared under the same conditions, but lacking in both type and constitutional vigor. The weaker bird is lacking in width and length of back. She also cuts off triangularly in both breast and body, which lessens capacity and weakens constitution.

Fig. 6. In this male, note particularly the **compactness** and **strength of conformation**.

Fig. 7. **A weak male**, loosely built and **of poor** conformation.

Fig. 8. A Rhode Island Red pullet of standard type and vigor. Egg production, 197.

Fig. 9. A Rhode Island Red pullet lacking in standard type and vigor. Egg production, 51.

Hens of the type shown in fig. 9 have neither the constitution nor the capacity necessary for high egg production. Fig. 10 shows the difference in width between the weaker and the stronger birds. Fig. 10 also shows the timid, less active and cowardly disposition of the weaker bird. The stronger and more vigorous birds have greater capacity, which is measured in finger widths and the distance from the end of the breast bone to the pelvic bones. The good type pullet shown in fig. 10 measures five fingers in capacity and not quite three fingers in width between the pelvic bones, while the lower vitality bird has a capacity of less than three fingers and a width between the pelvic bones of less than two fingers, a very marked difference.

Fig. 10. These birds illustrate the difference in width of breast and body of strong and weak females.

Figs. 11 and 12. These two figures, with figs. 13 and 14, show the improvement in the type of male birds between 1915 and 1920, due to selection and the introduction of new blood.

Figs. 13 and 14. These figures, with figs. 11 and 12, show the improvement in the type of male birds between 1915 and 1920, due to selection and introduction of new blood.

STANDARD COLOR

The standard color of the Rhode Island Red, described by the Amercian Standard of Perfection, in the male is "rich, brilliant red," in the female, "rich even red," except in sections where black is specified. Rhode Island Red breeders and judges differ some in the interpretation of what is rich, brilliant red.

The defects that were most common in the Rhode Island Reds that we had to work with were an excess of black in sections specified to be red. Slate and smut under-color were found in some birds and also an occasional out-cropping of white over the kidneys and in the base of the hackle of male birds, and frequent gray in the wings. A great many of the females showed mealiness and shafting in the breast and some of them black ticking on the wing bows.

While the females that we worked with carried a great many defects in color, they had excellent egg production. We tried to eliminate as many of the defects in color as possible and maintain the high egg production. Different matings were made in an attempt to find what produced an even rich red and what caused the different defects.

In those individuals that had the richest red color there was often an excess of black. The problem was to eliminate the black pigment and retain the rich brilliant red. In closely studying the make-up of the individuals, it was found that many of them showed a black pigment that was very black and intermingled with red. Where the black was found it appeared as a distinct bar of slate or black pigment in the backs of either the females or the males, while below this line of black pigment was a rich red under-color. These individuals proved to be by far the better birds in transmitting rich red color.

It was not found desirable to mate males with females when both carried black pigment in those sections specified to be free from black. Better results were obtained where the males were free from excess black and mated with females carrying a distinct bar of black pigment in the under-color of the back. Where the black pigment faded out or smoked towards the base of the feather, the brilliant red color was not transmitted.

Most of the individuals that showed a smoky under-color had a tendency to throw white, either in the neck, the wings or over the back, in the male birds. It appeared that the smoky or grayish under-color had an apparent mingling of white pigment or absence of red pigment, appearing white in the offspring.

A common breeding practice to secure a darker red color in the offspring is to use males carrying an excess of black with females light in color, figuring that the offspring will be darker in color. The offspring will carry a darker shade, but with a new defect equally as serious as light color — shafting, mealiness and mottling, produced by mating colors that are of an uneven shade. An even shade of red cannot be secured by mating dark with light. The best results in securing a rich, even shade of red were secured when females with the same shade of red as the male were mated.

In mating the best test as to shade of color can be made if the back of the female is matched with the breast color of the male. The more brilliant and nearer the same shade the undercolor and the surface color are, the better birds they will be. A larger percent of the offspring were free from color defects where neither the males nor females carried an excess of black.

However, it is impossible to have very many individuals that do not carry some defects. The problem is to mate them so that the differences will be less noticeable in the offspring.

COLOR MATING SUGGESTIONS

1. Have both males and females of the same shade of red color and free from excess black and white.
2. Mealiness, shafting and mottled condition of the plumage comes from mating males and females not of the same shade of red color.
3. White in the base of the hackle and over the kidneys is quite apt to come from smoky under-colored females.
4. A distinct bar of slate in the under-color of females, if mated with a male free from excess black, should not be considered objectionable.
5. A male carrying no black ticking in the hackle, with a few distinct narrow bars of slate in the back, should not prove objectionable if mated with hens free from excess black.
6. Females with ticking on the wing bows should not be mated with males carrying a bar of slate or ticking in the hackle.
7. It is not good practice to use males carrying excess black and rich red plumage on very light colored females.

EGG PRODUCTION

The trap nest was used to obtain the egg production of the females used in the experiment. Three types of producers are found. They have been classified as high producers, medium producers, and low producers.

High Producers. The egg record of Hen No. 1408 (cover fig.) shows the monthly egg production over a period of five years. The record of another high producer, No. 322, shows what usually happens in the second and third years, a drop of 73 eggs the second year and 17 the third year. Hen No. 25,227 has a rather unusual record. She did not get started to laying until December 6 her first laying year. For some reason she was thrown out of production in January. In her second laying year she came back with an exceptionally heavy egg production. Her broodiness did not increase the second year. A monthly egg production of from 25 to 28 eggs is usually found in the best hens.

EGG RECORD OF HEN NO. 1408

Col. 9-21 Breed S. C. R. I. R.

Year	Pen No.	Nov.	Dec.	Jan.	Feb.	Mar.	Apr.	May	June	July	Aug.	Sept.	Oct.	Total
1916-17	Tile 4	18	24	23	17	16	26	24	20	6	15	28	21	238
1917-18	Lot 7	13	0	0	5	24	19	29	27	29	25	26	10	207
1918-19	Pen 4	6	0	0	3	21	27	29	25	25	9	27	26	198
1919-20	Col. V	15	0	0	0	16	24	24	10	26	25	24	19	183
1920-21	Col. IX	5	0	0	0	18	22	24	8	0	18	22	5	122

EGG RECORD OF HEN NO. 322

Col. 9-21 Breed S. C. R. I. R.

Year	Nov.	Dec.	Jan.	Feb.	Mar.	Apr.	May	June	July	Aug.	Sept.	Oct.	Total
1918-19	13	22	21	19	25	23	22	20	19	20	15	5	224
1919-20	0	0	9	4	15	22	23	22	19	21	12	4	151
1920-21	0	0	7	13	20	19	19	23	13	19	1	0	134

EGG RECORD OF HEN NO. 25227

Col. 9-21 Breed R. I. Red Pullet

Year	Nov.	Dec.	Jan.	Feb.	Mar.	Apr.	May	June	July	Aug.	Sept.	Oct.	Total
1919-20		20	0	21	23	26	26	13	10	9	8	20	176
1920-21	17	2	1	13	26	26	23	25	22	25	17	10	207

Medium Producers. In the class of medium producers, the egg record of Hen No. 376 shows a total of 138 eggs. She did not get started to laying the early part of November, due to later maturity, and stopped laying in September her first two years, but in her third year she carried her production thruout the fall months but at a slow rate of speed.

EGG RECORD OF HEN NO. 376

Col. 13 Breed R. I. Red

Year	Nov.	Dec.	Jan.	Feb.	Mar.	Apr.	May	June	July	Aug.	Sept.	Oct.	Total
1918-19	0	10	22	12	10	26	15	B 9	B 19	12	3	0	138
1919-20	0	0	0	1	21	13	19	12	B 10	5	0	0	81
1920-21	0	0	0	0	3	15	13	6	11	8	14	11	92

Hen No. 379 shows more clearly the type of producer laying over a long period of time, but at a slow rate of speed. Her

highest month production was 21 eggs, while the best month for Hen No. 1408 was 29 eggs in May, the same month that Hen No. 379 laid her 21 eggs.

Hen No. 392 shows a higher rate of speed, a shorter laying period, and, probably due to lack of constitution, she is unable to make a good egg record.

EGG RECORD OF HEN NO. 379

Col. 13, Pen 4   Breed R. I. Red

Year	Nov.	Dec.	Jan.	Feb.	Mar.	Apr.	May	June	July	Aug.	Sept.	Oct.	Total
1918-19	0	6	5	1	15	15	20	16	18	10	8	12	127
1919-20	0	0	0	8	13	16	21	4	9	9	9	0	89
1920-21		0	0	2	15	6	6	11	4	18	4	0	66

EGG RECORD OF HEN NO. 392

Pen 4 L. H., Col. 16   Breed R. I. Red Pullet

Year	Nov.	Dec.	Jan.	Feb.	Mar.	Apr.	May	June	July	Aug.	Sept.	Oct.	Total
1918-19		0	0	21	25	24	2	24	B 14	10	0	0	120
1919-20		0	0	0	4	19	3	4	B 17	11	13	0	71
1920-21		0	0	0	1	8	11	0	18	B 11	14	0	63

Poor Producers. A study of the records of Hens No. 346 and No. 25,247 shows clearly the low rate of speed. In the record of Hen No. 25,247, notice the large number of months, both at the beginning of the year and at the end of the year, when no eggs are produced.

EGG RECORD OF HEN NO. 346 (See fig. 5)

Col. 13   Breed S. C. R. I. R.

Year	Nov.	Dec.	Jan.	Feb.	Mar.	Apr.	May	June	July	Aug.	Sept.	Oct.	Total
1918-19	0	0	0	1	4	17	B 17	0	5	5	5	0	54
1919-20	0	0	0	12	20	13	13	B 7	3	8	0	0	76
1920-21		0	0	8	3	11	8	8	2	0	1	0	41

EGG RECORD OF HEN NO. 25247

Col. 13   Breed R. I. R. Pullet

Year	Nov.	Dec.	Jan.	Feb.	Mar.	Apr.	May	June	July	Aug.	Sept.	Oct.	Total
1919-20	0	0	0	0	15	7	18	17	B 5	0	0	0	62
1920-21	0	0	0	5	16	6	12	0	0	0	0	0	39

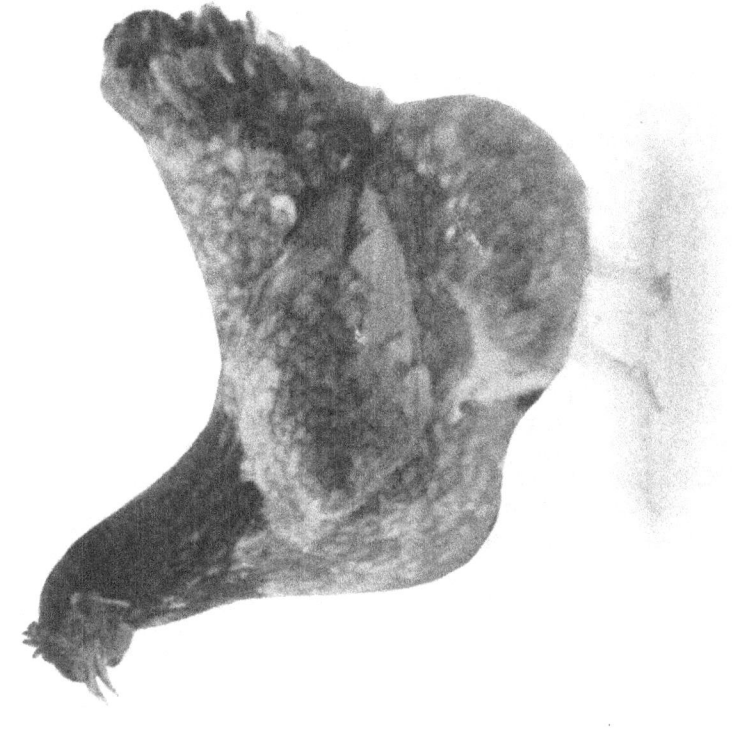

Fig. 15. Hen No. 2338, one of the original hens with an egg record of 256 eggs in her first laying year.

Fig. 16. Hen 337, daughter of No. 2338, with a record of 221 eggs.

Fig. 18. Hen No. 25023, a granddaughter of fig. 15 (No. 2338) and a daughter of fig. 16 (No. 337). Egg record, 224.

Fig. 17. Hen 335, another daughter of No. 2338 out of the male shown in fig. 13.

Fig. 19. Hen 27760, out of hen in fig. 16, from male shown in fig. 14.

Fig. 20. No. 2152, with a high producing record interrupted by broodiness once during each of the summer months.

BASIS OF SELECTION FOR EGG PRODUCTION

Type and constitutional vigor, as previously discussed, are the first characteristics that are demanded for egg production. High egg production was credited to a hen that laid 150 eggs or more; medium producers laid between 100 and 150 eggs. Low egg production was less than 100 eggs.

The outstanding characteristics of high-producing hens were noted particularly in a clean-cut head that showed quality, with alert, prominent eyes, and no tendency toward meatiness or puffiness. The appearance of the head in the high-producing birds is very clearly shown in the accompanying figures. It was found that there is a distinct difference in the high-vitality heads. Some of the birds showed strength, vigor and constitution, but were coarse, heavy and meaty, and there was a decided tendency for these birds to be low or medium producers. Constitutional vigor must be had in the high-producing birds, but coarseness, meatiness and a head approaching masculinity should be discriminated against. The clean-cut, sharp featured, feminine head is very necessary for high egg production.

The low-producing birds showed a tendency to fall into two groups, one of low producers that were small in size, lacking in vigor, constitution and capacity, and another group of large, heavy, fat, with slow rate of speed. These birds were coarse, heavy, thick and meaty in the head and thick and meaty in the pelvic bones, and had a tendency to become fat and broken down. Some of them made a good production the first year, but did not produce heavily the second or third year.

METHOD OF SELECTING WITHOUT THE USE OF THE TRAP NEST

Low and medium-producing hens can be eliminated from the flock, either during the late fall months or just before the breeding season starts. High-producing Rhode Island Reds will carry their plumage until late September, October, or even early November, before molting, while those hens of less production will molt earlier in the summer. The later molting hens can be marked with colored spiral bands and used the next year in the breeding pen. The pullets can be selected at the time of coming into egg production, provided they are hatched so as to become mature before December first. Feeding and management will have a great deal to do with the growth of the pullets. Pullets that did not lay before March 1 were found to be invariably low producers and most of them culls. All pullets that laid at a high rate of speed during January and February were found to be good producers. The speed of production was determined by the conformation of the body, the capacity, the thinness and straightness of the pelvic bones, the pliability of the abdomen and the loss of yellow pigment in the beak and shanks. Pullets that show qualities of high egg production may be used for breeding if they possess a strong constitution.

A further selection of the hens that were marked in September and October as late molters should be made again in February. The better producers should be laying or coming into egg production. The high-producing birds will show a quicker and more rapid molt than will the poorer producers, which will show a longer period of rest and a slower molt.

BROODINESS

Broodiness has not appeared to be a serious hindrance to egg production in our Rhode Island Reds. Some of our highest producing birds have been broody a great many times, but their speed of production before going broody and after the broody period has apparently been increased. It was found that it is necessary to be very particular in the handling of the broody hens, or the egg production is affected. The loss of time was increased if the hens were not placed in the broody coop the first night that they remain on the nest. Plenty of the same kind of feed that they were in the habit of getting, with an ample supply of fresh water, apparently shortened the broody period and brought them back into laying more quickly than where a special method of feeding was used.

A YEAR'S EGG RECORD OF A HIGH PRODUCER (fig. 20), WHICH WAS BROODY ONCE EACH SUMMER MONTH

N	Dec.	Jan.	Feb.	Mar.	Apr.	May.	June	July	Aug.	Sept.	Oct.	Total
						B	B	B	B	B		
25	21	22	16	28	24	20	15	18	13	18	13	239

Heavy egg production is secured by selecting those individuals that have the capacity to take care of a large amount of food and house the reproductive system. The individuals must also have vigor and vitality to stand the strain of heavy production.

A good layer should have three important characteristics, winter egg production, high speed of production and ability to lay in the summer and late fall months. The hen shown on the cover page shows the type of hen not only capable of laying a number of eggs, but of maintaining high egg production over a period of years. Not only does this hen lay a large number of eggs, but her hatchability is also good. It has been found n this experiment that there was no direct correlation between high egg production and low hatchability. Some of the highest-producing hens showed the greatest hatchability, while many of the poorer producers showed lower hatchability.

www.ingramcontent.com/pod-product-compliance
Lightning Source LLC
Chambersburg PA
CBHW062208220526
45470CB00009B/2967